THE CAVENDISH SQUARE ANIMAL LIBRARY

The World of SHARKS

BY JENNIFER LOMBARDO

Cavendish Square

Published in 2026 by Cavendish Square Publishing, LLC
2544 Clinton Street Buffalo, NY 14224

Website: cavendishsq.com

Cataloging-in-Publication Data

Names: Lombardo, Jennifer.
Title: The world of sharks / Jennifer Lombardo.
Description: Buffalo, NY : Cavendish Square Publishing, 2026. | Series: The Cavendish Square animal library | Includes glossary and index.
Identifiers: ISBN 9781502675217 (pbk.) | ISBN 9781502675224 (library bound) | ISBN 9781502675231 (ebook)
Subjects: LCSH: Sharks–Juvenile literature.
Classification: LCC QL638.9 L596 2026 | DDC 597.3–dc23

Editor: Jennifer Lombardo
Designer: Deanna Lepovich

The photographs in this book are used by permission and through the courtesy of: Cover (image) Shane Myers Photography/Shutterstock.com; cover and various pages (wave vector) Ball Apirak/Shutterstock.com; various pages (tape element) Tamara Luiza/Shutterstock.com; p. 4 Jsegalexplore/Shutterstock.com; p. 6 LuisMiguelEstevez/Shutterstock.com; p. 8 Jonas Gruhlke/Shutterstock.com; p. 9 mauritius images GmbH/Alamy Stock Photo; p. 11 (top) Mark_Kostich/Shutterstock.com; p. 11 (graph: sharks) Tom Fricker Design/Shutterstock.com; p. 11 (graph: person) Michal Sanca/Shutterstock.com; p. 12 9bdesign/Shutterstock.com; p. 13 (main) Simon Burt/shutterstock.com; pp. 13 (inset), 20, 26 Alessandro De Maddalena/Shutterstock.com; p. 14 Izen Kai/Shutterstock.com; p. 16 frantisekhojdysz/Shutterstock.com; p. 18 Sara Nadeea/Shutterstock.com; p. 19 Ruben Martinez Barricarte/Shutterstock.com; p. 22 Barbara Ash/Shutterstock.com; p. 24 Peter Douglas Clark/Shutterstock.com; p. 28 (top) Simon Mustoe/Shutterstock.com; p. 28 (bottom) Michael Bogner/Shutterstock.com; p. 29 wildestanimal/Shutterstock.com; p. 30 (top) Henner Damke/Shutterstock.com; p. 30 (bottom) Isogomphodon oxyrhynchus (Müller & Henle, 1839) 2432753577/Wikimedia Commons; p. 32 jose amaral/Shutterstock.com; p. 33 Rich Carey/Shutterstock.com; p. 34 This_is_JiHun_Lee/Shutterstock.com; p. 35 NOAA/NCEI World Ocean Database/NASA; p. 37 Allexxandar/Shutterstock.com; p. 38 IlluminatiXXX/Shutterstock.com; p. 40 Irina Kvyatkovskaya/Shutterstock.com; p. 41 Pocket shark noaa 1/Wikimedia Commons.

CPSIA compliance information: Batch #CSCSQ26: For further information contact Cavendish Square Publishing LLC at 1-877-980-4450.

Printed in the United States of America

CONTENTS

INTRODUCTION

When people think of sharks, the first that comes to mind is often the great white. This iconic shark is the kind most often featured in movies, including *Jaws*, *The Little Mermaid*, *Finding Nemo*, and *Aquaman*. However, there are more than 500 different species of sharks, all just as amazing as the great white!

Sharks come in many different sizes, from the 7.9-inch (20 cm) dwarf lanternshark to the 40-foot (12.1 m) whale shark. Their looks also vary widely; some of the most unique-looking animals on Earth are sharks, including the goblin shark, the hammerhead shark, and the frilled shark.

No matter what they look like, all sharks have skeletons made of cartilage instead of bone. This tough, bendable tissue, which is also what human ears and noses are made of, makes sharks light, fast, and **flexible**. This helps them swim faster than other fish, allowing them to catch up to their prey. The cartilage in their jaw also gives them their powerful bite. Cartilage's flexibility lets a shark open its mouth very wide, so its downward bite is fast and hard. Think of how hard you can clap your hands when they are only a few inches apart, versus when they are brought together from as far apart as you can stretch your arms. This is similar to how hard a person can bite with their jaw made of bone, versus how hard a shark can bite with their jaw made of cartilage.

Although many people are scared of sharks, only 12 species are known to be dangerous to humans. Most shark attacks on humans are believed to be accidents, since sharks often stop attacking after the first bite. This may be because the sharks mistake humans for their usual prey, such as seals. Experts say sharks have much more to fear from humans than the other way around.

Sharks are fierce fish with unique features, and there is a whole world of them waiting to be explored!

Angel sharks look like the ocean floor, which allows them to hide. They can hide even more by covering themselves with sand.

CHAPTER 1

SHARK FEATURES

Most sharks have long, **streamlined** bodies that are wider at the front than they are at the rear. This shape lets them swim for long periods of time without getting tired. However, there is some variation in shark bodies when adaptations are needed. For example, sharks such as the spotted wobbegong and the angel stay close to the ocean floor. These sharks have wider bodies than sharks that swim through open water. The angel shark looks similar to a manta ray, while the spotted wobbegong looks a bit like a flattened-out catfish. These two shark species often lie partly buried in sand or mud, waiting for prey to swim by. They are camouflaged to blend in with the ocean floor.

THAT'S WILD!

Wobbegongs have a very wide mouth, and they'll eat just about anything that can fit in it. If an animal is too large to swallow, the shark will chomp down with its teeth and hold the prey there until it gets tired and dies.

JUST KEEP SWIMMING

A shark is a fish, but its body is different than that of most fish. One big difference is that, unlike most fish, sharks do not have an organ called a swim bladder. The swim bladder helps a fish float as well as move around in the water: The fish can take air into its swim bladder to rise toward the surface, or release it to help it sink deeper in the water.

A shark's light cartilage skeleton helps it float. So does its large liver, which has a lot of oil in it. One species of shark, called the sand tiger, has a special adaptation. It can swallow air and hold it in its stomach, which then acts like a swim bladder. This allows the sand tiger shark to stay still and float in the water. Other sharks must swim to avoid sinking.

Some sharks also need to swim to stay alive. The great white, whale, and mako sharks are some examples of species

All sharks are fish. This means they breathe using gills that take in oxygen from water.

> **THAT'S WILD!**
> The thresher shark uses its unusually long tail to catch prey. It can whip its tail up to 80 miles (129 kmr) per hour, slapping and stunning a fish so it can't swim out of reach.

that use obligate ram ventilation. This means they breathe by moving. As they swim, water flows over their gills, allowing them to take in oxygen. The faster they move, the more oxygen they get. If they stop moving, they get no oxygen, and they will die.

Most sharks, however, use a combination of obligate ram ventilation and buccal pumping to get oxygen. This means that when the shark is not moving, it uses its buccal, or cheek, muscles to pull water into its mouth and over its gills. Many sharks that take in oxygen this way have a special organ called a spiracle that helps them pull water over their gills.

FINS AND SKINS

Fins help a shark balance, swim, and float. The fin that is shaped like a triangle in the middle of a shark's back is called the dorsal fin. Dorsal fins keep sharks from rolling. Every shark also has a pair of fins that stick out sideways.

All sharks have a dorsal fin, but some – such as the one on the frilled shark, which lives in very deep water and is rarely seen by humans - are so small that they're easy to overlook.

These fins are called pectoral fins. They help keep a shark from sinking as it swims.

> **THAT'S WILD!**
> Most sharks are brown, gray, or olive green. One exception is the blue shark. Its back is a deep blue color, and its stomach is white. This helps it blend into the water from both above and below.

A swimming shark can come to a stop by tilting its pectoral fins. Many sharks that live on the sea floor use their pectoral fins to help them crawl forward along the ground. A shark's tail fin also helps it move. Swaying its tail fin gives the shark a strong push.

> **THAT'S WILD!**
> Unlike most sharks, whale sharks do not have a **nictitating membrane.** Scientists think the dermal denticles on their eyes are there to protect the eyes.

A shark's skin is covered with thousands of small, spiky "teeth." These point in the direction of the shark's tail, and they help reduce the friction from the water as the shark swims through it, allowing the shark to swim faster. Scientists call the "teeth" on sharks' skin placoid scales, or dermal denticles. They make the skin feel just like sandpaper. Whale sharks even have dermal denticles on their eyeballs!

WHAT BIG TEETH YOU HAVE!

A shark's teeth are shaped to handle the kind of food it eats. Great whites and many others have pointed lower teeth and triangular, **serrated** upper teeth. This combination helps them hold onto and rip apart large prey, such as sea lions and dolphins. Some sharks, such as mako sharks, have thin, needle-like teeth for holding on to smaller, slippery fish. These teeth can be straight or curved, depending on the shark species.

Some bottom-dwelling sharks eat shellfish. Their front teeth are pointed to help them grip their prey, and their back teeth are wide and flat for crushing hard shells.

EXTINCT SHARKS

Sharks have existed for hundreds of millions of years. In fact, they appeared on Earth before trees did! Many species of sharks have gone extinct over the years. Cartilage does not fossilize the way bone does, so most of what scientists know about extinct sharks comes from their teeth. This is how scientists know about the megalodon – the largest shark to ever exist in history, as far as we know.

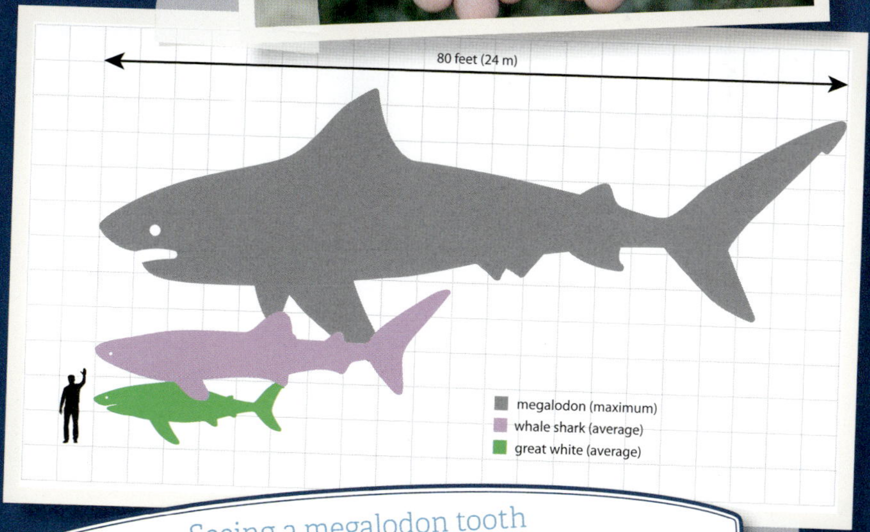

Seeing a megalodon tooth (*above left*) next to a great white shark tooth (*above right*) gives an idea of its size. Experts believe the megalodon could reach up to 80 feet (24 m) in length, and its jaw was wide enough to swallow two adults side by side.

The goblin shark does not swim as fast as some of its prey. To help the shark catch food, it has a special adaptation: It can push its jaw several inches out of its mouth!

Some shark teeth are non-functioning, or **vestigial**, teeth (*right*). Species such as basking (*above*) and whale sharks swim slowly with their mouths open so they can catch tiny creatures, such as plankton. They don't need teeth, so they have this kind.

A shark's teeth grow in rows. Depending on the species, a shark can have between 50 and 300 teeth in its mouth at one time, arranged in rows. When one falls out, it is replaced by one from the next row. Some sharks can go through more than 30,000 teeth in their lifetime.

Hammerhead sharks have a wide field of view because their eyes are so far apart. They can spot prey moving on either side of them. As the hammerhead swings its head from side to side, it can even see what is behind it!

CHAPTER 2

SHARK SENSES AND ABILITIES

Sharks have seven senses, compared to humans' five. Many also have special abilities that help them catch prey or protect themselves in some way. For example, the scalloped hammerhead shark dives to the ocean floor to catch its prey. Because sharks are cold-blooded, diving that deep can make a shark too cold to keep moving. To stay warm, the scalloped hammerhead closes its gills while it's diving, which prevents heat from escaping its body.

THAT'S WILD!

If there is not a lot of food available, some sharks can go for weeks without eating. The longest a shark has ever been observed going without food is 15 months.

SIGHT

Sharks cannot see colors or pick up on fine detail from more than about 50 feet (15 m) away from a target. For these reasons, there is a myth that sharks have bad eyesight. In reality, their eyes are very sharp. Although they can't tell the difference between different colors, they can pick up on color contrasts. Light doesn't travel well underwater, so even seeing 50 feet (15 m) is a sign of amazing eyesight. The average human

This tiger shark may look like it's rolling its eyes, but the white part is actually its nictitating membrane. Sharks can raise or lower this membrane whenever they need to, just like people can close their eyes.

THAT'S WILD!

For a long time, people believed sharks were attracted to the color yellow. Recent studies have shown that it is easier for sharks to see yellow in water because of the high contrast between yellow and the blue water.

can't see more than 5 feet (16 m) underwater. To help them even more, sharks' eyes are set on the sides of their head, allowing them to see all around them. Like cats, they also have a layer of reflective cells in their eyes to help them see better in low light. This is why sharks' eyes sometimes look like they are glowing in pictures.

SMELL

A shark's sense of smell is even better than its sense of sight. In fact, for many sharks, this sense is their strongest. Sharks use their noses to locate prey, tell friends apart from enemies, and find a mate. Some sharks have a sense of smell so keen that they can sniff out one part of blood per one million parts of water. That's the equivalent of one teaspoon of blood in an average-sized swimming pool.

THAT'S WILD!

Great whites, tiger sharks, and bull sharks are the three species most likely to attack humans. Bull sharks are particularly **aggressive** and often swim in shallow water, where humans are more likely to be, which makes them the most dangerous species to humans.

TASTE

A shark's sense of taste does not help it find food, but experts believe it can help them decide if the food they've found is worth eating. Many shark attacks on humans stop after the shark has taken one bite. This could be because their sense of taste tells them they've found something that is not part of their natural diet. Scientists are still learning more about sharks' sense of taste.

TOUCH

Sharks have very sensitive skin, and some have extra organs that improve their sense of touch. For example, some species have barbels, which are fleshy **appendages** near the mouth. They are more prominent on bottom-feeding species, such as nurse sharks, because the barbels help

This close-up image of a bamboo shark shows its two barbels.

these species feel along the ground for prey that is hidden under the sand.

Sharks can even feel things with their teeth, which have pressure-sensitive nerves in them. Some experts suggest that shark attacks that end after one bite are the result of a shark using its teeth to investigate what's in front of it, not realizing the damage its bite can do.

HEARING

A shark's hearing is very sensitive. They can track sounds from more than 800 feet (243 m) away. It helps that sound travels farther through water than it does through the air, but this is still an impressive feat. Sharks are especially good at picking up on low sounds. Their ears are two small holes on the sides of their head, behind the eyes.

A shark's ears are easy to miss. One is circled in this photograph.

EXTRA SENSES

Sharks have two more senses than humans. One is their ability to sense **electromagnetic** fields. This sense is called electroreception, and sharks do it with special organs called ampullae of Lorenzini. From a distance, these look like tiny dots all around a shark's nose and lower jaw. Depending on

This picture of a shortfin mako shark clearly shows its ampullae of Lorenzini – the small black dots all over the shark's nose.

IT'S ELECTRIC!

Detecting electricity helps sharks find prey. This is because all living things give off a low-level electromagnetic field. Cells use electricity to send messages to each other. This is part of what keeps living things alive. Sharks can sense this electricity through their ampullae of Lorenzini and follow the electromagnetic current toward the organism, or living thing, that is giving it off.

The earth itself also gives off an electromagnetic field. This is why compasses always point north. Sharks can sense the direction of the earth's magnetic field and use it to figure out which way to go when the water is too dark or cloudy for them to see clearly. This helps them migrate over long distances.

the species, the number of ampullae of Lorenzini can range from a few hundred to more than 1,500.

A shark's second special sense comes from its lateral lines. These are made up of several small organs working together, similar to the ampullae of Lorenzini. However, rather than detecting electromagnetic fields, the lateral lines sense vibrations and pressure changes in the water, such as those made by prey moving nearby. Lateral lines are another way sharks sense prey when they can't see or hear it.

THAT'S WILD!

Scientists had no idea what the ampullae of Lorenzini did until 1971, although they had noticed as early as the 1930s that sharks reacted to electric currents.

Shark eggs are protected by tough cases known as "mermaid's purses." One example is shown here.

CHAPTER 3

A SHARK'S LIFE

With so many different species of sharks, it should come as no surprise that sharks have various diets, habitats, behaviors, and more. Even the way sharks are born is not consistent. Although they are not mammals, most sharks give birth to live young. Only about 25 percent of shark species lay eggs.

GIVING BIRTH

Some kinds of sharks, such as carpet sharks, lay eggs on the ocean floor. Shark babies, called pups, hatch from these eggs without any help from their mother. The egg cases come in different styles. Some have strings that wind around seaweed to keep them from being moved by the ocean currents into unsafe waters. Others are shaped like screws so they can be wedged into cracks, where they will stick tight until they hatch.

THAT'S WILD!

Shark pregnancies last at least nine months. Some species stay pregnant for more than a year. The frilled shark can have a **gestation period** of three and a half years – the longest of any shark as well as any vertebrate on Earth.

This pregnant nurse shark's pups are hatching from eggs inside her body. After about six months, she will be ready to give birth.

Many kinds of sharks, such as tiger sharks, keep their eggs inside their bodies instead of laying them.

These eggs have thinner, softer coverings. The pups from these eggs hatch while they are inside their mother. They stay inside their mother and continue to grow until they are fully developed. Some sharks, such as blacktip reef sharks, don't make egg cases at all. The pups grow inside the mother until they are ready to be born, just like human babies.

WHAT'S FOR DINNER?

A shark's diet can include many different creatures. Turtles, seals, jellyfish, squid, dolphins, plankton, and even other sharks can all be prey to various species. Some sharks hunt by waiting quietly and then grabbing a fish as it swims by. Many hunt while swimming and go after their prey in bursts of speed.

A shark often swallows its prey whole. However, if the prey is too big, the shark breaks it into smaller chunks. It can do this by biting the prey into pieces or holding the animal in its mouth and shaking it.

THAT'S WILD!

Sharks aren't just predators. They can also be prey for animals such as orcas, giant squids, and – in very rare cases – crocodiles.

One shark, called the cookiecutter shark, is only about 22 inches (56 cm) long, but it feeds on prey that is up to 10 times its size. The shark is bioluminescent, or naturally glowing, everywhere except for a small band around its neck, which researchers call the "dog collar." To animals such as dolphins, seals, and whales, this dog collar looks like a very small fish that would make a tasty snack.

When one of these animals swims down to catch what it thinks is its prey, the roles are suddenly reversed, and the predator becomes prey for the small shark. The cookiecutter shark lunges toward the larger animal and latches on with its small, pointy top teeth. Then it rotates its body and uses its larger bottom teeth to take a circular chunk out of its prey,

THAT'S WILD!

When large amounts of food suddenly appear in one place, such as when people throw a lot of bait into the water, it can trigger a feeding frenzy in some sharks. This is when the sharks start biting anything that moves, including each other.

The light circles on this whale's back are healed scars from cookiecutter shark attacks.

leaving a circular mark that looks like it was cut out with a cookie cutter.

FEW FRIENDS

Most sharks are loners. Males and females pair up to mate, but they don't form families. Mothers typically do not take care of their pups after they're born. However, a few types of sharks sometimes form groups. Great white sharks form packs to hunt for seals. Hundreds of scalloped hammerheads will join up to form a school during the day, but split up at night to hunt on their own.

Different types of sharks usually stay out of one another's territory, but sharks that live around coral reefs generally share their habitat. One shark species may hunt only by day, while another species may hunt only at night. Some species stay in shallow water. Others hunt in deeper water. This way, the species are not competing with each other for food.

THAT'S WILD!

Grey reef sharks have been seen hunting in packs with whitetip reef sharks, but not cooperatively. The whitetips are better at getting into coral reefs, so the greys follow them and eat the fish the whitetips chase out of the reef.

MIGRATION

Many animals live in one place for part of the year, then travel somewhere else to live for the rest of the year. This movement is called migration. Scientists have found that many sharks migrate. The reason for the migration typically depends on the shark species, but there are four main reasons.

The most common reason for sharks to migrate is to follow a food source. When prey migrates, sharks migrate with them. Sharks also migrate so that they can always be in warm water, since they are cold-blooded animals. A few species migrate to find mates. Male and female sharks do not always live in

A SYMBIOTIC RELATIONSHIP

Although sharks do not often form alliances with other sharks, many species of sharks have developed **symbiotic** relationships with some kinds of fish, such as remoras and pilot fish. The fish eat parasites off a shark's skin. This keeps the shark healthy, because parasites, which feed on the body of the shark, can cause serious illness, and even death if left long enough.

In return, the shark protects the fish from other predators and allows them to eat some of the food it catches. Some fish even clean the shark's teeth by picking bits of leftover food out from between them.

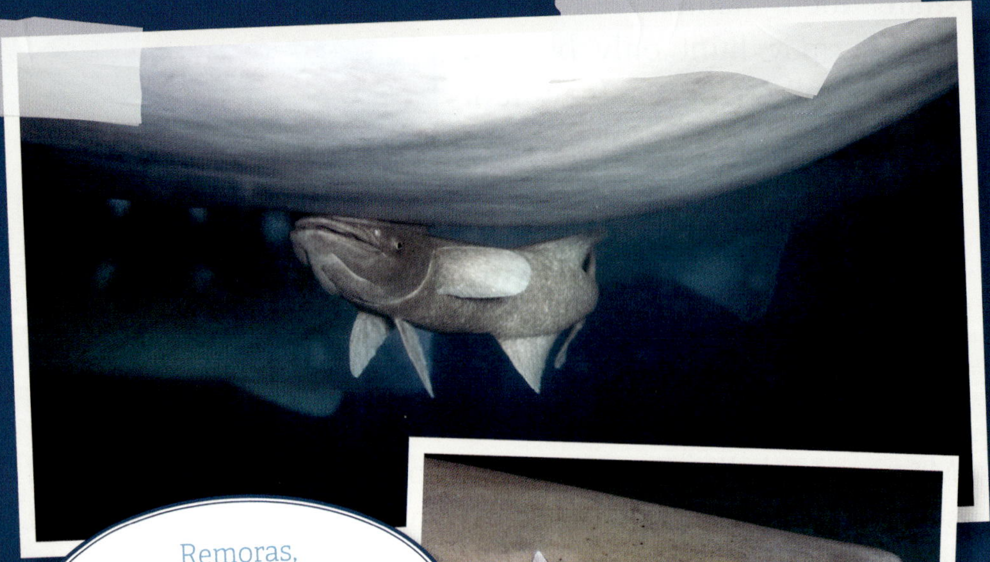

Remoras, shown here attached to sharks, are also called suckerfish because they use their dorsal fin as a suction cup to attach to a shark. Sharks sometimes slow down to give remoras a chance to attach.

From May to July, millions of sardines migrate along the coast of South Africa. To sharks such as this bronze whaler, this is like an all-you-can-eat buffet, and they're willing to travel to take advantage of it.

the same area, so they migrate to certain areas to meet up. Finally, many species migrate to specific parts of the ocean that are safe and comfortable enough for them to give birth.

The daggernose and sand tiger sharks are listed as critically endangered, which means they are highly at risk of dying out. The last time a daggernose was seen in the wild was 2016.

CHAPTER 4

SAVE THE SHARKS

Although sharks have been around for millions of years, they are now starting to disappear because of human activity. Experts say that more than 31 percent of all shark species are at risk of extinction, or dying out permanently. The biggest reason for this is overfishing. Sharks reproduce slowly, and high demand for shark products means that they are killed faster than they can repopulate.

Many species of sharks are sought after for things such as their meat, the oil in their liver, and their tough skin. Shark fin soup is a popular dish in some parts of the world, and it is not uncommon for fishers to cut the dorsal fin off and then throw the shark back into the water to die. This practice is called "finning."

Even when fishers are trying to catch other fish, sharks sometimes get caught in their nets. Other current problems that endanger sharks include climate change and pollution.

THAT'S WILD!

Shark fin soup was invented in China between 960 and 1279 CE. Shark fins were expensive, so serving this soup at a party was a way for people to show off how rich they were.

Shown here are dorsal fins from various sharks drying in the sun.

THE IMPACTS OF POLLUTION

Pollution has a major impact on all sea creatures, including sharks. Many objects used by humans end up in the oceans, either by accident or because they are dumped there on purpose. Until the 1970s, it was common for individuals and businesses to dump chemicals and garbage in the ocean without realizing how much harm they were doing. The effects of that are still being felt today, and while dumping has slowed down thanks to new environmental laws, people still do it illegally.

THEY ATE WHAT?!

Humans don't just throw plastic in the ocean. Some strange things have made their way into the water and then into a shark's stomach. These include:

- **license plates:** Tiger sharks are especially well-known for eating license plates. Some experts believe this is because the shiny plates reflect light, just like certain fish do.
- **a suit of armor:** One great white shark was found with an entire suit of armor in its stomach in the 16th century.
- **a chicken coop:** An entire coop, complete with the bodies of chickens, was found in the stomach of a tiger shark off the coast of Florida.
- **a bag of money:** Experts believe a robbery was committed in Florida and the robbers either dumped or lost their loot in the water, and it then ended up inside a shark.
- **fur coats:** Sharks are likely attracted to fur coats because they look like furry seals floating through the water. However, it is suprising that multiple fur coats have ended up in the path of sharks!

Plastic bags are common pieces of ocean pollution. It is especially easy for whale and basking sharks to eat them because they feed by floating through the water with their large mouths open.

Shown here is some garbage that has washed up on the shore. There is much, much more still in the ocean.

Pollution harms sharks in several ways. One is that toxic chemicals build up in their bodies, causing illness or death. Pregnant sharks can also pass on these chemicals to their pups. Additionally, sharks sometimes eat plastic items, which can get stuck in their digestive system and kill them.

CLIMATE CHANGE

Humans are polluting the air as well as the water. Smoke and chemicals from cars, factories, and more enter the atmosphere and trap heat, warming the planet up. This global warming leads to climate change, or a change in the long-term weather of the planet.

As the planet gets warmer, so do the oceans. Scientists measure heat in units called joules. A 100-watt light bulb uses 360,000 joules of energy every hour. The amount of heat we are putting into the oceans is so huge that it must be measured in zettajoules. One zettajoule is equal to one vigintillion joules—that's a 1 followed by 21 zeroes! This affects fish that need cool water to survive. As those fish die off, it becomes harder for sharks to find food. Sometimes this means sharks starve to death. Other times, it means they migrate to places they do not traditionally belong, disrupting the ecosystem in those places. If the ocean gets too warm, sharks may lose the ability to regulate their body

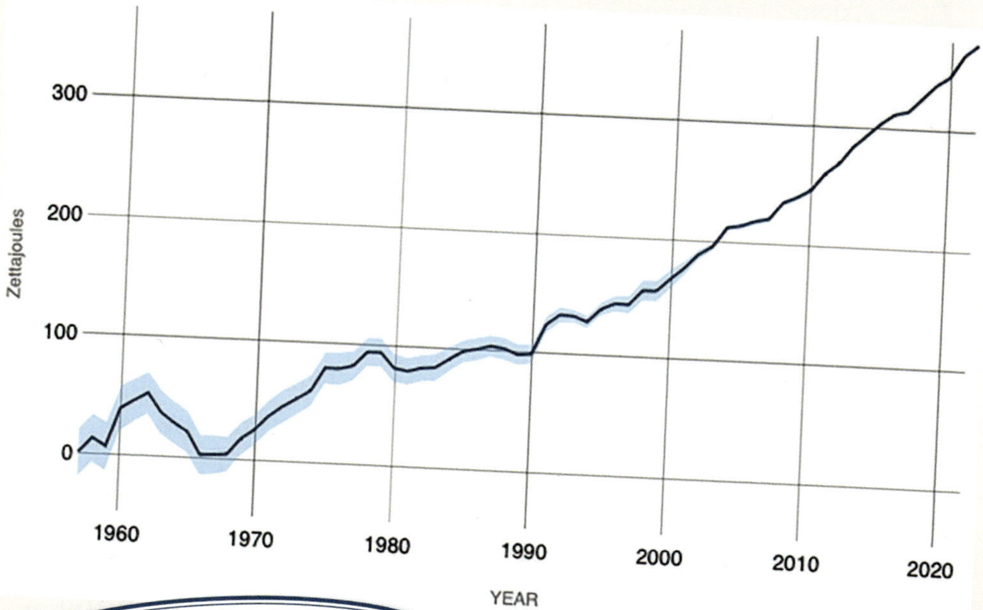

This chart shows the increase in ocean temperature since the late 1950s.

temperature. When they get too warm near the surface, they dive deeper to cool down. If the water does not get cold until they pass the point where they can comfortably dive, certain sharks will get sick or die because they cannot maintain the proper body temperature.

WHY SAVE THE SHARKS?

Since most people will never encounter a shark in their daily life, it can be hard for them to understand why it matters if some shark species go extinct. To them, the world seems the same whether there are sharks in the ocean or not. However, sharks play an important role in their environment, and changes in one part of the natural world affect everything else over time.

> **THAT'S WILD!**
>
> Sharks help increase the **biodiversity** of coral reefs by keeping down the population of large fish that prey on smaller fish.

One important role of sharks is that they protect coral reefs. Reefs form when animals called corals attach themselves to the ocean floor. Algae grows in the corals, and microorganisms that live there use the corals' waste products to perform photosynthesis. Because oxygen is a waste product of photosynthesis, coral reefs are an important source of oxygen as well as a home to many kinds of animals. Many species of sharks feed on fish that eat corals. If these sharks died out, the coral-eating fish populations would grow, and the corals could die. The presence of sharks helps keep the environment in balance.

Another way sharks help the world is by indirectly helping global economies. Shark meat, skin, and fins may bring a profit, but sharks can generate much more money when they're alive than when they're dead. This is because people often enjoy seeing sharks and other animals in their natural

The corals that make up reefs may look like plants, but they're actually animals! Sharks help protect them from fish that feed on them.

THAT'S WILD!

Sometimes sharks are killed because people who see them fear being attacked. However, sharks kill only about 10 people per year. In the United States, more people die from exposure to the cold, lawn mower accidents, and being struck by lightning.

Because so few sharks are dangerous to humans, swimming with them can be a fun and memorable experience.

habitats. Swimmers, snorkelers, boaters, and sightseers travel to places where sharks are known to live. They pay money to rent gear, hire tour guides, and take boat trips, hoping to see living sharks swim past. They also pay for things such as food and hotels while they are on their trip. All of this money helps the people who live in these areas.

A third reason to protect sharks is that they are living creatures that deserve to be left in peace. Finning is considered cruel because it kills sharks in a painful way. Sharks may not be as cuddly as cats or as cute as bunnies, but they are still animals that can feel pain.

THAT'S WILD!

Shark finning is outlawed in many places, but illegal hunters called poachers continue the practice anyway. It is the third highest illegal trade in the world, after drugs and weapons.

CONSERVATION EFFORTS

Once people began to realize how important sharks are to the world, they started passing laws to help protect them. For example, in 2000, the United States made finning illegal. More than 70 other countries also ban the practice. In addition, rules have been made to shorten fishing seasons in order to reduce the number of sharks trapped in nets.

Although some of these conservation laws have helped, many people look for loopholes, or ways to get around following them. For example, the U.S. law that outlawed finning was worded in an odd way: It said that fishers who were caught with a shark fin but no shark body would be punished. Fishers got around this by catching some sharks for their meat and others for their fins. To close the loophole, a new law was passed in 2010 that made it illegal for fishers to come ashore with shark fins that were not attached to the shark's body.

People and organizations around the world are working hard to save all the different species of sharks. At the same

HOW CAN YOU HELP?

Even if you aren't a marine biologist or a lawmaker, you can do your part to help save the sharks. Some options include:

- reducing your use of plastic to reduce the amount of pollution that ends up in the ocean
- taking a trip to see sharks to show local economies that living sharks are more valuable than dead ones
- adopting a shark through an organization such as Shark Angels
- checking to make sure that none of your lotions, makeup, or deodorants contain an ingredient called squalene or squalane, which comes from shark liver oil

Squalene or squalane is in many skin care products. Plant-based versions are better for sharks.

Researching sharks sometimes leads to new discoveries. This tiny shark, known as a pocket shark, was found in 2010. In 2015, after much study, scientists declared that it was an entirely new species.

time, researchers are also continuing to study sharks. Learning more about sharks can help people be more interested in saving them.

In the words of nature journalist Richard Louv, "We cannot protect something we do not love, we cannot love what we do not know, and we cannot know what we do not see."

THINK IT THROUGH

1. Why do you think there are so many different species of sharks in the world?
2. How could obligate ram ventilation be helpful to sharks?
3. Why do you think sharks lose teeth so easily?
4. Which species of shark interested you the most? Why?
5. Can you think of other ways to help save the sharks?

GLOSSARY

aggressive: Showing a readiness to attack.

appendages: Projecting parts of an animal or plant body.

biodiversity: The different kinds of life in an environment shown by numbers of different kinds of plants and animals.

electromagnetic: Relating to the magnetic field that's produced by a current of electricity.

flexible: Able to bend easily.

gestation period: The amount of time it takes a baby to be ready to be born.

nictitating membrane: A thin, transparent or translucent membrane found in many animals at the inner angle or beneath the lower lid of the eye and capable of extending across the eyeball.

serrated: Having a saw-toothed edge.

streamlined: Shaped to make travel underwater quicker and easier.

symbiotic: Relating to symbiosis, which is a close physical association between two organisms that are not similar to each other and that benefits each organism in different ways.

vestigial: Remaining in a form that is small or imperfectly developed and not able to function.

FIND OUT MORE

Books

Forster, Miriam. *Sharks: A Mighty Bite-y History*. New York, NY: Abrams Books for Young Readers, 2022.

Gish, Melissa. *Sharks*. Mankato, MN: Creative Education and Creative Paperbacks, 2024.

Jackson, Carlee. *Sharks: What Do Great Whites, Hammerheads, and Whale Sharks Get Up to All Day?* New York, NY: Neon Squid, 2022.

National Geographic Society. *Sharks: 300 Fin-nomenal Facts to Dive Into*. Washington, DC: National Geographic, 2023.

Ventura, Marne. *Sharks*. Minnesota, MN: Early Encyclopedias, 2023.

Websites

Kiddle: Shark Facts for Kids

kids.kiddle.co/Shark

Read more about sharks and take a look at some cool photos.

***National Geographic Kids*: Surprising Sharks**

kids.nationalgeographic.com/animals/fish/article/surprising-sharks

Learn more amazing facts about some species of sharks.

Shark Evolution: A 450 Million Year Timeline

www.nhm.ac.uk/discover/shark-evolution-a-450-million-year-timeline.html

London's Natural History Museum breaks down what we know about shark evolution.

Shark Facts for Kids

sharkfactsforkids.weebly.com/types-of-sharks.html

This website contains basic information on more than 30 different species of sharks, including basking sharks, hound sharks, and silky sharks.

Sharks4Kids

www.sharks4kids.com

This website contains facts, activities, and ways to help save the world's sharks.

Publisher's note to educators and parents: Our editors have carefully reviewed these websites to ensure that they are suitable for students. Many websites change frequently, however, and we cannot guarantee that a site's future contents will continue to meet our high standards of quality and educational value. Be advised that students should be closely supervised whenever they access the Internet.

INDEX

ABOUT THE AUTHOR

Jennifer Lombardo earned her BA in English from the University at Buffalo and still resides in Buffalo, New York, with her cat, Chip. She has helped write a number of books for young adults, on topics ranging from world history to body image. In her spare time, she enjoys cross-stitching, hiking, and volunteering with Habitat for Humanity.